Know how to solve radical + rational inequalities with a calculator b/c can't do algebra.

Take any rational func in form $y = a\sqrt{b(x-h)} + k$ + graph w/out calc.

3 = WP
1 = radical eq
1 = rational eq
1 = graphing
22 = radical + graph - no calc

$2\sqrt{(x-3)}$

WP - Mixing
Area
Dist

2 4 - 14

Know how to solve algebraically rational + radical equations + check answer.

pg 139

1, 8, 19, 24, 23, 30, 31, 46-49

241. 48, 49
243 62-63
252 38-43
263 44-51

262 263

26 - 28

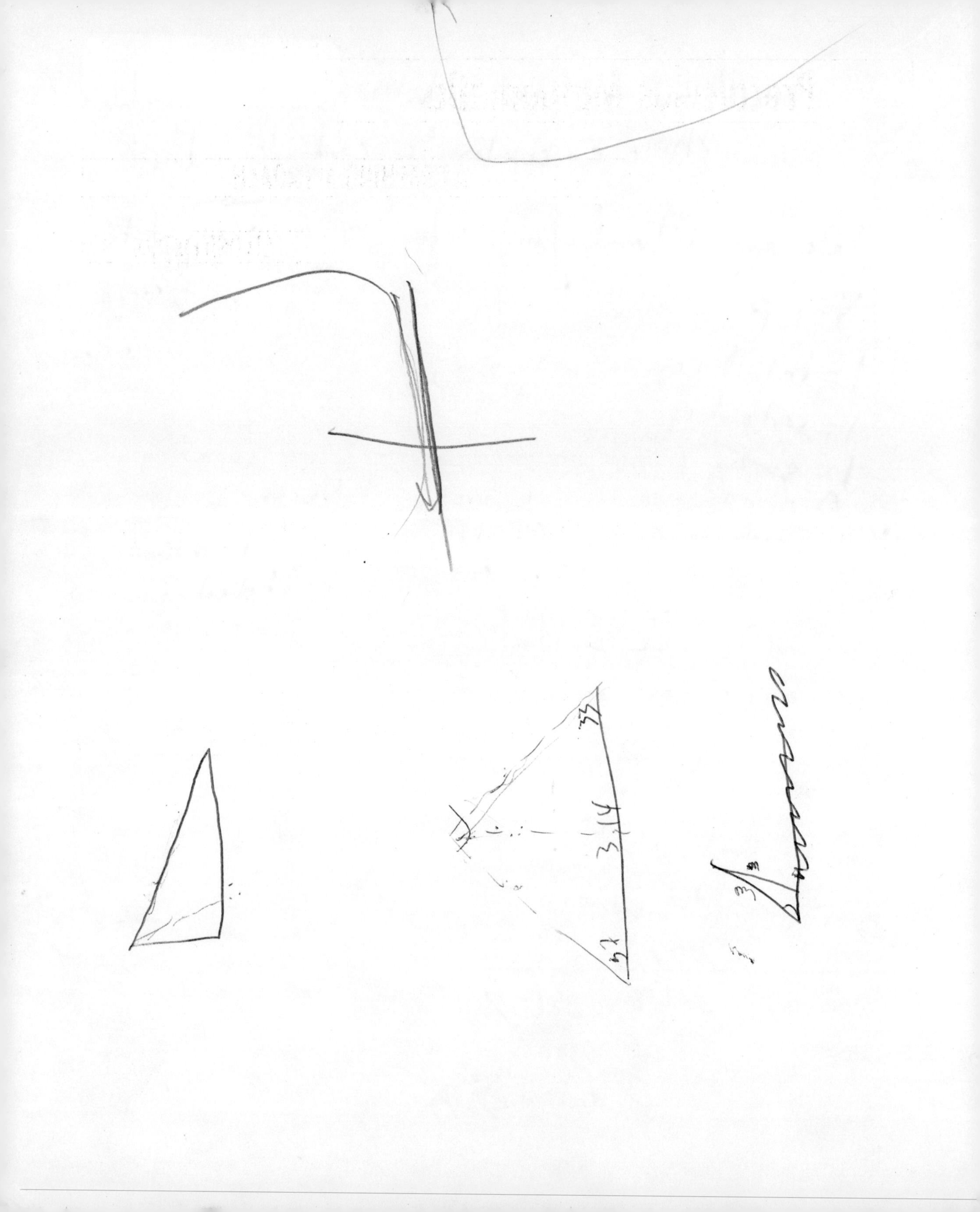

Precalculus Mathematics

A GRAPHING APPROACH

SECOND EDITION

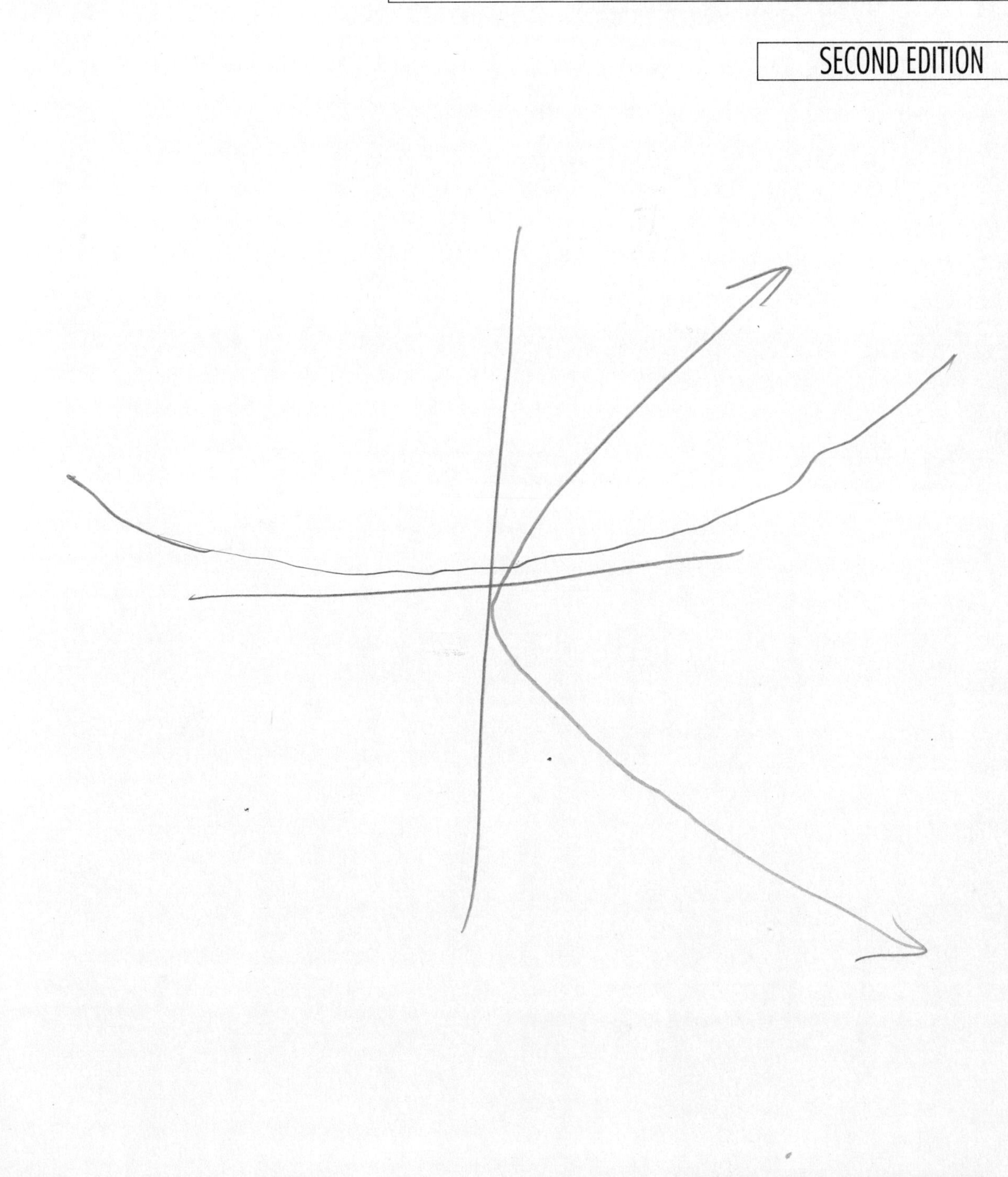